Gray Whale Migration

by Grace Hansen

Abdo Kids Jumbo is an Imprint of Abdo Kids
abdobooks.com

abdobooks.com

Published by Abdo Kids, a division of ABDO, P.O. Box 398166, Minneapolis, Minnesota 55439.
Copyright © 2021 by Abdo Consulting Group, Inc. International copyrights reserved in all countries.
No part of this book may be reproduced in any form without written permission from the publisher.
Abdo Kids Jumbo™ is a trademark and logo of Abdo Kids.

Printed in the United States of America, North Mankato, Minnesota.

052020

092020

THIS BOOK CONTAINS
RECYCLED MATERIALS

Photo Credits: Alamy, BluePlanet Archive, iStock, Minden Pictures, Shutterstock

Production Contributors: Teddy Borth, Jennie Forsberg, Grace Hansen
Design Contributors: Dorothy Toth, Pakou Moua

Library of Congress Control Number: 2019956491
Publisher's Cataloging-in-Publication Data

Names: Hansen, Grace, author.
Title: Gray whale migration / by Grace Hansen
Description: Minneapolis, Minnesota : Abdo Kids, 2021 | Series: Animal migration | Includes online
    resources and index.
Identifiers: ISBN 9781098202330 (lib. bdg.) | ISBN 9781098203313 (ebook) | ISBN 9781098203801
    (Read-to-Me ebook)
Subjects: LCSH: Gray whale--Juvenile literature. | Gray whale--Behavior--Juvenile literature. | Animal
    migration--Juvenile literature. | Animal migration--Climatic factors--Juvenile literature.
Classification: DDC 599.522--dc23

Table of Contents

Gray Whales

Gray whales live in the Pacific and Atlantic Oceans. One large population lives in the eastern Pacific Ocean.

South for the Winter

Eastern Pacific gray whales live off the coast of Alaska. There are around 24,000 individuals in the group. Each October, these whales start **migrating** south toward Mexico.

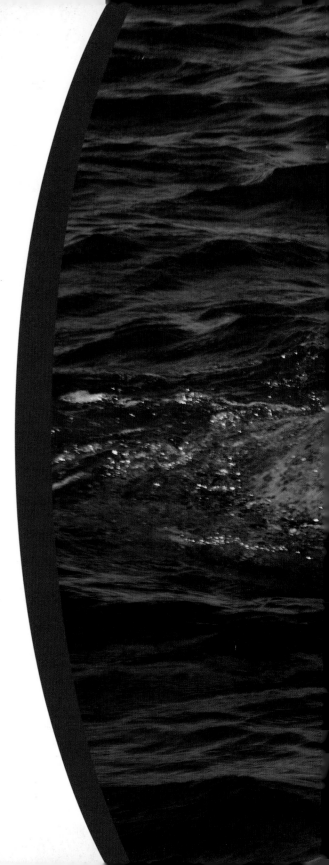

Gray whales can **migrate** around 75 miles (121 km) a day. In one year, they will travel about 13,000 miles (21,000 km). This is believed to be the longest migration of any **mammal**.

8

The whales make it to Mexico by late December. The water is warmer off the coast of Mexico. It is the perfect place to have young.

Pregnant females find shallow

lagoons. The lagoons are warm.

They are also safe from sharks

and other **predators**.

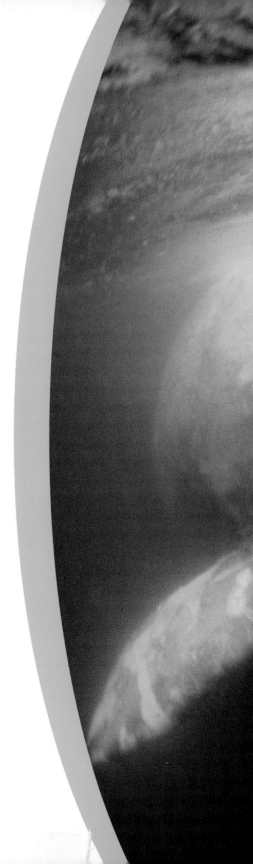

Females give birth to one calf at a time. Calves are around 13 feet (4 m) long when they are born. They rely on their mothers for milk.

Back Home to Eat

In springtime, the migration home to Alaska begins. Males and newly pregnant females are the first to move north.

New mothers and calves stay for a month or two longer. Calves need to gain lots of **blubber** before the long journey.

It is important to make it back to Alaska. This is where the gray whales eat. They must gain lots of weight to prepare for the next **migration** south.

21

Gray Whale Migration Route

● Summer Home　● Winter Home　– – – – Route

Glossary

blubber – the layer of fat beneath the skin of whales and other large sea mammals.

lagoon – a shallow body of salt water by the sea.

mammal – a warm-blooded animal with fur or hair on its skin and a skeleton inside its body. Mammal mothers make milk to feed their young.

migrate – to move from one place to another for food, weather, or other important reasons.

predator – an animal that hunts other animals for food.

23

Index

Abdo Kids ONLINE
FREE! ONLINE MULTIMEDIA RESOURCES

Visit **abdokids.com** to access crafts, games, videos, and more!

Use Abdo Kids code

AGK2330

or scan this QR code!